What

M000077320

Are you willing to gamble with your eternity?

For the skeptic, Dr. Vanarthos brings to the discussion of who Jesus really is, the staggering - and highly compelling - mathematical evidence supporting His claim to be Messiah - God's promised savior for all men. Be careful if you choose to read it - there is little argument against the proofs he offers.

Brian P. Gilman, CLM, CPA

Dr. Vanarthos is a passionate believer and a compassionate man. He brings both traits to bear in this book by engaging the reader with a gentle conversational style, yet presenting his argument with a vigor and enthusiasm that keeps the reader engrossed. Furthermore, he knows his subject exceedingly well, and backs up every claim with multiple layers of examples. As a bonus, the reader will no doubt enjoy Dr. Vanarthos' sense of humor and originality as he presents concepts in ways that the reader may have not have heard before. This book will not only reinforce your belief; you will want to share it with others – especially those you know who are *on the fence*.

Philip R. Saba, MD

What are the Odds is compelling and convincing. If I weren't already a believer, I *would* be after reading this book! Dr. Vanarthos does a masterful job making something so complex and vast so simple. I will want to share this with every skeptic patient I see.

M. Alan Dickens, MD

It is hard to fully embrace "truth" unless it has first been questioned. Dr. Vanarthos had serious doubts about Christianity. His journey from skeptic to Christ follower, based, in part, on the information contained in this book, is very compelling and will cause every reader to consider the validity of Christianity.

Rev. Dr. Todd von Helms
Chaplain, St. David's School
Raleigh, NC

What Are The Odds?

Are You Willing To Gamble With Your Eternity?

William J. Vanarthos, MD

Author: William J. Vanarthos
Editor: Greg McElveen
Cover Design: Olivia Vanarthos and Josh Cianca
Illustrations: Copyright © 2016 William & Olivia Vanarthos

All Scripture verses are taken from the New American Standard translation of the Bible unless otherwise specified.

Library of Congress Control Number: 2016945225
Library of Congress subject headings:
1. BS647-649   Prophecy
2. BT1095-1255   Apologetics. Evidences of Christianity
3. BS2415-2417   The Teachings of Jesus
4. BT750-811   Salvation. Soteriology

BIASC Classification Suggestions:
1. REL006140 RELIGION / Biblical Studies / Prophecy
2. MAT029000 MATHEMATICS / Probability & Statistics / General
3. REL067030 RELIGION / Christian Theology / Apologetics

Paperback ISBN-13:   978-1-937355-36-4
eBook ISBN-13:   978-1-937355-37-1

Published by Big Mac Publishers
www.bigmacpublishers.com / Kingston, TN 37763
Written and processed in the United States of America

Table of Contents

Dedication

*This book is dedicated first and foremost
to my Lord and Savior, Jesus Christ,
Who saw fit to open my eyes
to the truth of His Word.
It is my prayer that He will use
the content of this book to impact
the lives, and perhaps even
the eternal destinies, of its readers.
It is also dedicated to my wife, Jill,
and my daughters, Olivia and Madalyn,
for their encouragement and support
in making this dream a reality.*

"But what about you?" Jesus asked.

"Who do you say that I am?"

-Matthew 6:15
(Berean Study Bible)

Preface

Despite having attended church throughout much of my elementary and secondary school years, it was during my college education and medical training that I began to question the validity of the Bible, and entertain the notion that it was either a myth or simply a collection of outdated and irrelevant stories. It was not until I completed medical school and was in residency training that God captured my attention and changed my life.

Since then I have been studying the Bible seriously for nearly three decades and have become aware of statistical probabilities that shocked me and proved undeniable. What I discovered might surprise you, too.

How would you respond to these questions?

- What are your beliefs regarding Jesus?
    - Who is He to you?
    - Is He simply a good moral teacher?
    - What are the chances you are wrong?
- How confident are you of what awaits you after death?

The compelling illustrations contained in this book may impact your viewpoint forever.

Introduction

Don't you just love *CliffsNotes*? I do. These student study guides explain and summarize literary and other works in condensed pamphlet form or online. I think it's so great that people on their staff took the time to read a mammoth work like Fyodor Dostoevsky's *The Brothers Karamazov* and summarized each chapter in about a page and a half. That's awesome!

For those of you who watched television in the 1950's and 60's (or reruns in the 70's), remember Sergeant Joe Friday, the detective from the TV drama *Dragnet*? His signature line was "Just the facts, ma'am." I loved that guy. Don't get me wrong; details are extremely important. Ask any criminal lawyer or tax accountant. But sometimes we tend to get bogged down in the details at the expense of the big picture.

Regarding the subject of this book, the big picture is what counts most. That's why this book is as short as it is…to focus on "just the facts…," if you will. It is written primarily for the skeptical and often science-minded individuals like me who may too easily dismiss the claims of the Bible, in general, and Jesus, in particular; but my hope is that it also becomes a tool for all people who may not take time to attend church or study the Bible.

I'm not concerned with arguing the existence or origins of God, intelligent design, evil, morality, etc. There is a plethora of comprehensive books on the subject of Christian apologetics that line the shelves of libraries and bookstores, and populate the internet. They are certainly more thorough than this one (there is a list of some of my favorites in Appendix A, along with other recommended reading). This book is not intended to be a tome. I want it to be short enough to feasibly read in one sitting, and small enough to carry comfortably in a pocket or purse. Hopefully, then, it can easily be shared with friends and family.

This book is a compilation of statistics based on extreme odds that not only lend credibility to the Bible's and Jesus' claims, but are virtually indisputable, and help form the foundation of the Christian faith. I have tried to state them in a clear and compelling fashion with the hope (and even the expectation) that you will embrace them, so that you may know the truth.

Please take this short journey. I believe you'll be glad you did.

William J. Vanarthos, MD

Chapter 1: No Apologies

Jesus Christ is Lord.

I mention that right off the bat for several reasons:

1. It sums up the most important truth that I will try to prove in this book.
2. I want to be transparent with you.
3. If after reading this book you disagree with that statement, then I would genuinely be interested in hearing your **proof** to the contrary.

It might surprise you to learn that, to date, I have never met **anyone** who took seriously the challenge to prove otherwise and did not become a follower of Christ. I am not talking about the person who reads, "Jesus Christ is Lord" and thinks, "I don't believe it." Rather, I am referring to serious students who are concerned about the question of their eternal destiny to the point that they won't rest until they are convinced of the truth. "That's not possible," one might say, "to know with certainty what is truth." I disagree. You see, I believe there can only be one truth regarding the deity of Christ. After all, we can't all be right, considering that so many people's views are in direct opposition to one another. And that one truth *can* be known—in fact, that is God's chief desire.

So, who is the last one standing when all the smoke clears? I believe it is the Christian. And by Christian, I don't mean the hypocrite who gives lip service to Christ on Sunday mornings and lives immorally the rest of the week. Nor do I mean one who adorns himself or herself with Christian symbols, yet curses God's name...or even someone who simply acknowledges that Jesus lived. I am talking about the person who believes the gospel message of the Bible.

What is this message? In Greek, "gospel" is translated "good news." So, what is this good news?

In 1 Corinthians 15:1-4 the Apostle Paul sums it up by stating,

> "Now I make known to you brethren [brothers], *the gospel* [emphasis mine] which I preached to you, which also you received, in which also you stand, by which also you are saved, if you hold fast the word which I preached to you, unless you believed in vain. For I delivered to you as of first importance what I also received, that *Christ died for our sins according to the Scriptures, and that He was buried, and that He was raised on the third day according to the Scriptures . . .*" [emphasis mine]

The Bible plainly states that we are all sinners (Romans 3:23). This simply means that we have all offended God in some way and have broken His law. It matters not what the offense is. Many of us like to rank sins, one being worse than another, but not so with God. To Him, sin is sin, and equally offensive.

Generally, sins fall into one of three major categories; namely, the lust of the flesh, the lust of the eyes, and the pride of life (1 John 1:16), to which we have all fallen prey. However, even "lesser" infractions such as white lies, exaggeration for our own benefit, gossip, and impure thoughts qualify us as sinners. And, because of our sin, we are separated from God (Isaiah 59:2), are spiritually dead (Romans 6:23, Ephesians 2:1-3), cannot please God (Romans 3:10-12), and will suffer damnation (2 Thessalonians 1:9).

As much as we would all like it to be true, a holy God cannot wink at sin. He *must* judge it. This makes perfect sense, does it not? In life, we all want justice when it is *we* who are offended or a heinous act has been perpetrated against another. Yet, when it is we who are the offenders, we want God to look the other way. But, like a fair and wise judge in a court of law, the Supreme Judge of the universe cannot and should not turn a blind eye.

Many of us believe that, somehow, we can earn our way into Heaven and restore our relationship with God by working harder. There is a common misconception that if we do more good than bad, then we tip the balances in our favor and God will grade us on the curve. But the Bible says that we are incapable of removing the guilt of our sinfulness through our own efforts. Galatians 2:21 says, ". . . if righteousness comes through the Law [the dos and don'ts of moral behavior], then Christ died needlessly." Sin is our infection and our ultimate demise. You see, we cannot become righteous by what we do. We are dead in our sins (Romans 5:8). It doesn't take a physician to know that cardiopulmonary resuscitation (CPR) does not work on a corpse. Dead is dead.

This all sounds bad. So, what's the *good* news? Simply this: ***Since we cannot remove our own sins, God must do it...and He did!*** He bore upon Himself the penalty for our sins by coming to earth in the form of a man, namely Jesus Christ (John 1:1, 14, Colossians 2:9), and subjecting Himself to the horrors of crucifixion on a Roman cross (1 Peter 2:24). Why? Because He loves you. And by raising Jesus to life, God conquered the last great barrier, death, so that it might not have its fearful grip on our lives.

It's not complicated. God loves you and provided a way for you to be reconciled to Him through His Son Jesus (John 14:6, Acts 4:12). In return, He asks simply that you believe. Nothing more of us is required. ***That is good news!***

So, true Christians accept the ***unmerited favor*** of God (that's grace as the Bible defines it) by faith and place confidence in the fact that this sovereign God knows, even more than they, what is best for their lives. That's the real meaning behind knowing that "Jesus is Lord." And so they trust, obey and follow Him. Is that what you believe? If not, does that worry you or elicit a visceral reaction that makes your blood pressure go up just a little? If so, perhaps you're ready to consider what this little book offers.

"…and you will know the
truth and the truth
will make you free."
- John 8:32

Chapter 2: Shades of Gray

I'm glad you're still with me, and I sincerely pray that you and your loved ones will be with me someday in Heaven at the foot of God's throne. I hope the prospect of that idea thrills you. I believe we were all created *on* purpose *for* a purpose, but we will never fully understand what that purpose is until we believe that Jesus Christ is King of kings and Lord of lords. He alone is the Way, the Truth, and the Life (John 14:6).

I am a simple man. I like the notion of absolutes and "one way," and I like to think in terms of black and white. Unfortunately, it seems our society prefers to see in shades of gray. I should know. I have spent 25 years as a radiologist looking at shades of gray all day long. Don't get me wrong. I also like gray, especially since what hair I have left is becoming more so at an alarming rate. But some things really are just black and white.

I have come to appreciate that salvation in Jesus Christ is as black and white as it can possibly get. The leap of faith one must take to believe that notion is not across some giant chasm. Think of it more as stepping over a crack in the sidewalk. It doesn't take as much effort as you may think.

After completing her experiential pilgrimage and embracing Christianity, Nobel Prize winning novelist

Sigrid Unset risked the ridicule of the Scandinavian intelligentsia and concluded: "If you desire to know the truth about anything, you always run the risk of finding it." (1)

My intent in this book is to expose you to enough information so that you will find the truth and agree without question and without apology that "Jesus Christ is Lord."

Let's begin.

Chapter 3: Seeing is Believing

I graduated from college with a B.S. in Mathematics. Frankly, the degree was more a matter of convenience than it was of interest because the mathematics curriculum gave me flexibility to simultaneously fulfill the requirements for premed. However, one class I found fascinating was Probability & Statistics. I'm not sure why, except that some of us, I think, are just wired to take things more at face value if they seem reasonable (mathematically speaking).

Science-minded individuals, in general, live in a world of reasons and proofs. Research is based on repeatable, reliable experimentation. The challenge is "Prove it to me. Oh, and by the way…when you're done…prove it to me again." The all-to-familiar "I'll believe it when I see it" attitude is pervasive.

Some Christians I know like to flip things around and retort, "Well, believing is seeing!" That's clever, and there is truth to it, of course, since we are dealing with matters of faith. But, seeing *is* believing also. Let's face it. After witnessing a full-blown miracle, believing is a little easier. Then, the more one believes, the more one sees as God honors that belief and reveals yet more to appreciate. It is a cycle. Faith is deepened as one trusts in God more.

Saint Augustine is quoted as saying, "Faith is to believe what you do not yet see; the reward for this faith is to see what you believe." (2) I agree and hope that the information in this book will help you to believe, that you might see all that God intends for you to see.

Las Vegas rakes in its riches everyday based on odds and probability. On average, the house advantage of slot machines is around 10%, which means that, for every dollar you spend, casinos return 90 cents and keep a dime. Perhaps this is why slots are in the airport, bathrooms, etc. Soon they'll be in hotel showers...and people will play! Clearly, the odds are *not* in your favor. But if they were, you'd be rich. Here's the good news: the odds *are* in your favor when it comes to the riches that matter most. In fact, you stand to inherit so much more than the world can offer. Intrigued? Read on.

Chapter 4: The Bible

Sanctify them by the truth; Your Word is truth.
— John 17:17

Is the Bible reliable? Sadly, many have dismissed its credibility. They consider it an allegory at best, perhaps simply a fantasy or something worse. I suspect that those who feel that way may be relying on the media, commonly held traditions, or others' opinions instead of reading, studying, and researching the Bible for themselves. Is this true of you? On what is your belief based?

It's a paradox to me that people believe all kinds of crazy things rooted in magic, speculation, emotion, and opinion without investigating their origins and principles on which they are based. Look at horoscopes, for instance. Assuming that there are generally equal numbers of Pisces as there are Leos and Scorpios, etc., what are the odds that 1/12th of the United States population, or approximately 30,000,000 people, are having exactly the same kind of day?

Now let's consider these basic facts regarding the Bible:

1. It is a book. No one refutes that. Whether or not one considers it the Word of God is another story, but clearly it's a book that has been around for centuries.

2. The Bible has been critically read, studied, scrutinized, probed, dissected…you name it…by millions (billions) of people, yet has stood the test of time and remains the best-selling book of all time by a margin too vast to calculate.

3. Apart from a few stylistic variations, to my knowledge, ***it contains no contradictions of content***. For those of you who have been taught otherwise or, for some reason, have come to believe that the Bible is riddled with errors and discrepancies, please show me one…a ***legitimate*** one. "Errors" pointed out by skeptics usually derive from opinions, hearsay, misunderstandings, lack of understanding of the culture in which the books of the Bible were written, or trivial quibbles, and can be categorically dismissed.

Think about this for a moment. Here is a text composed of 66 separate "books" written over thousands of years by over 40 different authors from all walks of life, from different parts of the world and of all ages, and, yet, it reads like a seamless narrative from cover to cover. Try this little experiment. Grab 40 friends (or strangers) and have each write a sentence expressing one idea on a piece of paper without consulting one another. Now read the sentences aloud in any order. What are the odds that they will form a connected, coherent story? How about 10 friends or just 5? Rather than a full sentence, make it just one word. You get the point. Not only does the Bible make sense, but also portions of Scripture separately written centuries apart affirm and ratify one another.

In his classic work entitled, *"Jesus Christ: The Greatest Life...A Unique Blending of the Four Gospels,"* Johnston M. Cheney took 23 years to study the gospels (that is, the eyewitness accounts of Matthew, Mark, Luke, and John) and to blend them together. He did not exclude **any** word from **any** of the gospels, but exact words and phrases common to more than one gospel were printed only once. Amazingly, the four gospels, blended together, read as one story written by one author rather than four.

In writing their gospel accounts, Matthew, Mark, Luke, and John could never have orchestrated such a phenomenon, especially since they were written years apart, decades in the case of John's account. Clearly, this phenomenon of consistency would imply that the gospels (and, by deduction, all of Scripture) were supervised by One who is unaffected by time and space.

4. As stated in other encyclopedic and well-researched works*, the number of original manuscripts of the New Testament exceeds 20,000, far surpassing any other written documents by orders of magnitude. In fact, the *Iliad*, second to the New Testament in manuscript authority, has only 643 manuscripts in existence. (3) Does anyone challenge the authenticity of the *Iliad*?

5. The Bible's historical accounts have been substantiated and validated archeologically hundreds of times over the centuries. While archaeology does not, in itself, prove the Bible to be true, and while there are yet unresolved mysteries, to my knowledge, **no discovery has ever invalidated Biblical accounts**.

6. The Bible never attempts to hide harsh realities. It is not simply a collection of platitudes and moral adages. Unlike many other inspiring works, it does not shy away from the realities of the corrupt human condition. Virtually every

*Sources include such works as: *Unshakeable Foundations*, by Norman Geisler & Peter Bocchino, 2001, p. 256; *Christian Apologetics*, by Norman Geisler, 1976, p. 307; article *"Archaeology and History attest to the Reliability of the Bible,"* by Richard M. Fales, Ph.D., in *The Evidence Bible*, Compiled by Ray Comfort, 2001, p. 163; and *A Ready Defense*, by Josh Mcdowell, 1993, p. 45.

type of malicious, sinful behavior and transgression is exposed, including abortion, adultery, deception, blasphemy, incest, stealing, rape, murder, lying, fraud, gossip, coveting, idolatry, drugs, alcoholism, extortion, kidnapping, slander, witchcraft, and sexual perversions to name a few. Not to mention less conspicuous but equally perverse and damaging behaviors like shame, bitterness, arrogance, contempt, fear, hatred, disobedience, lust, malice, anxiety, and, of course, pride.

Did you know that much, if not most, of the Bible was written by murderers (Moses, David, and Paul)? Here's the point: if you or I were writing a religious document, we would likely paint ourselves in a favorable light. But God is real and used imperfect people to create and preserve a perfect document on which we can all rely.

Here's the biggie:

7. In short…fulfilled prophecy. The Bible is *unique* in that it foretells specific events years (and often centuries) in advance, which have been fulfilled in astonishing detail. Dr. Hugh Ross, in material issued through his ministry "Reasons to Believe," explicitly and scientifically showcases several fulfilled prophecies that are documented historically and, as he puts it, "exemplify the high degree of specificity, the range of projection, and/or the 'supernature' of the predicted events." (4) In other words, they could not be faked or manipulated.

They are truly fulfilled prophecies, not self-fulfilling prophecies.

Even in assigning conservative probabilities to the event or events that comprise the prophecies, Ross clearly demonstrates that the odds of these happening with such amazing accuracy are astronomical, and, for all intents and purposes, essentially impossible. Taken alone, some of the prophecies theoretically could happen; however, when combined with others, the probability of all prophecies being fulfilled becomes a statistical impossibility, even by scientists' standards. Though arbitrary, generally speaking, a statistical impossibility is one that has a less than 1 in 10^{50} chance of happening.

That's 1 out of
100000000000000000000000000000000000000000000000000.
Not good odds. (5,6)

Based on a collection of just 9 events highlighted in some of the prophetic books of the Old Testament (Jeremiah, Isaiah, Zechariah, Ezekiel, Micah, and Daniel), along with some historical events in the time of Israel's kings, Ross demonstrates through the principles of composite probability* that the odds of all 9 prophecies happening with the degree of accuracy spelled out in the Bible is approximately 1 in 10^{104}! (7)

* For an explanation of composite probability, see pages 33-34

That's 1 out of
1000000000000000000000000000000000000000000000000000000
00000000000000000000000000000000000000000000000000000000.
Not good odds at all!

This number far exceeds the currently accepted number of atoms in the *entire* universe (estimated at 10^{66*}) by a factor of more than 10^{35} and is akin to the odds of the second law of thermodynamics being reversed! (8)

Those are the odds for just 9 fulfilled prophecies. Bear in mind that the Bible confidently declares roughly 2500 prophecies, **over 2000 of which have already been fulfilled to the letter without error!** Wow! You may want to read that sentence again and let it sink in for a minute. People can't even accurately predict the weather with certainty or how their fantasy football star will do next week. And predictions regarding the New York Stock Exchange...forget about it!

Applying these odds in practical terms to get an idea of what we're talking about is, in itself, impossible, but some attempt will be made in the next chapter as we explore probabilities related to prophesy regarding Jesus since they number in the hundreds rather than thousands.

Here's something extra for science geeks. The Bible

* Some claim this number is closer to 10^{80}

is chock full of fundamentals of science which often contradicted the science of its day and years following the time in which they were written, but which have been proven accurate and substantiated by more modern scientific discoveries. For instance, the Old Testament is explicit in pointing out that:

- The earth is a sphere (Isaiah 40:22)
- The number of stars exceeds a billion (Jeremiah 33:22)
- Wind blows in cyclones (Ecclesiastes 1:6)
- Blood is a source of life and healing (Leviticus 17:11), amongst others (9)

It also comments on topics such as:

- The water cycle (Ecclesiastes 1:7, Isaiah 55:10)
- Effects of emotion on physical health (Proverbs 16:24 and 17:22)
- Control of contagious diseases (Leviticus 13:45-46 and in various passages throughout Leviticus, Numbers, and Deuteronomy)
- The importance of sanitation to health and food as it relates to the control of cancer and heart disease (10)

Surprised? I was, but no longer. It is compelling to me that an infinitely wise God, who knows all things, is the author of the Bible and simply used men as quill pens, if you will, to provide the truth about Himself and

the timeline of human history on which we all find ourselves.

In 2 Timothy 3:16 Paul writes, "All Scripture is inspired by God and profitable for teaching, for reproof,
for correction, for training in righteousness..." Though
he was referring specifically to the Hebrew Scriptures
(Old Testament) in the context in which he wrote, the
same is true of Paul's letters and the established canon
that comprise the New Testament. Indeed, the whole
Bible is infallible and as applicable today (*if not more
so*) as it was centuries ago when first recorded. This
phenomenon could only be accomplished by an eternal
God not bound by time and space.

As if all this wasn't enough to convince someone
that the Bible is supernatural and authentic, consider
Bible codes, which are only now being appreciated with
the advent of computers and complex software. There
are literally dozens of popular books and articles on the
topic, but allow me to emphasize a couple of examples
that are simply mind-blowing.

In his book, *What is the Truth? The Case for Biblical Integrity (Kindle Edition),* Dr. Chuck Missler points
out that in chapter 38 of the book of Genesis detailing
the account of incest involving Judah and his daughter-
in-law, the name of Boaz appears in 49-letter intervals.
That is, specifically, the Hebrew equivalent of the letters
B-O-A-Z are each separated by 49 letters. The same is

true of the names Ruth, Obed, Jesse, and David within the same chapter *in that order*. The significance of this phenomenon may not be apparent to readers who are unfamiliar with the Bible, but this list of names is the exact genealogy of King David mentioned in the book of Ruth, yet encoded in the book of Genesis centuries before any of these people were even born! The notion that this sequence of letters in perfectly spaced intervals occurs randomly is inconceivable. (11)

Next, consider heptadic structures. "Heptadic" means "of, or relating to a heptad" or simply "seven-fold." Missler also describes these in detail. Simply speaking, these make reference to the frequency with which the number 7 (symbolically associated with completion or perfection in the Bible) appears throughout the Scriptures, either overtly or structurally.

Taking, for example, the genealogy of Jesus Christ located in the opening verses of Matthew 1 as originally written, all of the following are divisible by **7**:

- the number of vowels
- the number of consonants
- the number of words that begin with a vowel
- the number of words that begin with a consonant
- the number of words used overall

- the number of words that occur more than once
- the number of words that occur in one form
- the number of words that occur in more than one form
- the number of nouns (and only 7 words are *not* nouns)
- the number of names overall
- the number of male names
- the number of generations (12)

That's remarkable, to say the least! Someone today would have a tough time figuring this out using an elaborate computer algorithm. Bear in mind that this was penned by a first century tax collector with little or no education. It seems difficult to argue that this wonder was not supernaturally inspired.

Dr. Ivan Panin, an emigrant from Russia, Christian mathematician and scholar, and 1882 Harvard graduate, discovered the heptadic structures in 1890. Dubbed "the father of Bible numerics," Panin published numerous articles demonstrating that the numerics upon which the Bible was constructed proves an elaborate design.

Among numerous remarkable discoveries in his extensive work on this subject he discovered the following: In the book of Matthew, there are 42 words (a multiple of 7) that are unique to his gospel, containing 126 letters (also a multiple of 7). This phenomenon is entirely possible if Matthew collaborated with his fellow gospel writers or wrote his gospel last. However, the unique vocabulary used in the gospel of Mark is also a multiple of 7, as are those of the gospels of Luke and John. It is unlikely that they somehow collaborated to do this (why would they?), and not all of these writers could possibly have written their works last. By the way, the same is true of the writings of James, Peter, Jude, and Paul. As Missler emphasizes, these are the fingerprints of God! (13)

Truly, it is hard to argue against the suggestion that *the Bible stands apart as a reliable source of truth and knowledge and bears unmistakable evidence of a supernatural imprint.* Indeed, I believe that nothing is more certain than God's Word.

Chapter 5: Jesus Christ

But these are written that you may believe that Jesus is the Messiah, the Son of God, and that by believing you may have life in his name.

— John 20:31

We now examine the main point of the book.

Jesus Christ is Lord.

For the sake of space and brevity, I will not detail Jesus' profound teachings, predictions, miracles, transfiguration, and the like to prove his Lordship. (These are all readily available in any of the four Gospel accounts found in the Bible, and, of course, I encourage you to read these!) However, I believe with only a few clear points His Lordship can be convincingly established.

In the Christian classic *Mere Christianity*, C. S. Lewis introduces the "Lord, liar, lunatic" concept, expounded upon by Josh McDowell (14) and multiple other Christian authors, to convey the only three sensible options in which one can classify Jesus Christ. Jesus distinguishes Himself from all other religious leaders that ever lived in that, not only did He alone proclaim to be the God of the universe in the flesh, but also that

He would die and rise from the grave. What a preposterous claim! Surely it is one that can easily be dismissed.

Following the framework of options Lewis defined as we consider this claim of Jesus, let's begin with two simple facts:

1. Jesus is a man who walked the earth in the first century. There is generally no real argument here; men of all religions, even most hard-core atheists, will acknowledge that. (Despite this general stipulation and the substantial historical record, remarkably there remain a few who completely reject the notion that Jesus ever existed. I will come back to that point shortly.)

2. Jesus claimed to be God. "Hold on a minute," you say. "He never really said He was God." Well, actually, yes He did…on multiple occasions and in multiple ways. One Scripture verse easily sums it up for me. After threatening to stone Jesus for His outrageous statement "I and the Father are one" (meaning that they are one and the same) recorded in John 10:30, Jesus asks His fellow Jews in verse 32, "I showed you many good works from the Father; for which of them are you stoning Me?" Here's the kicker (John 10:33): "The Jews answered Him, 'For a good

work we do not stone You, but for blasphemy; *and because You, being a man, make Yourself out to be God.*'" [emphasis mine] The theological Jewish brain trust on the great "I AM" (the name God designated for Himself in the Old Testament) bore evidence of His making this assertion on multiple occasions; it is what ultimately got Jesus crucified. *So, His own enemies provide clear confirmation that He made this claim!*

Having established that Jesus really did *claim* to be God, let's stop and apply some basic logic. He either is God or He isn't. I would imagine that many readers don't believe He is, so let's begin with the argument that He's *not* God in the flesh. Then, applying C.S. Lewis' criteria and logic, He would have to then fall into one of two categories; namely, liar or lunatic.

Consider the lunatic perspective. If Jesus was insane, He would not only have had to convince *Himself* He was sane, but also thousands of others including some who lived intimately with Him for several years. Of course, on the surface this option is entirely possible since it can also be said of other notorious egomaniacs, such as Jim Jones, who convinced his following of 909 inhabitants of Jonestown, Guyana (including 304 children) to drink cyanide-laced *Flavor Aid.* (15) However, unlike Jones and other delusional characters throughout the ages, there is no evidence to suggest that Jesus ever displayed any signs of mental instability.

On the contrary, He appeared to understand His identity and mission, and seemed to always be in control. Also, insane people typically say crazy things. Rather, Jesus' teachings were masterful (effectively using parables and common objects of the day to make a point), awe-inspiring, and challenging, often leaving the crowds astonished (Matthew 7:28-29) and even impressing the religious elite of His day (John 7:45-46) who wanted nothing more than to trip Him up and prove Him wrong. Unlike the rabbis of His day who (albeit impressively) mechanically and habitually reiterated the Scriptures from memory, Jesus brought fresh, new revelation and insight causing people and leaders to acknowledge, "Never has a man spoken the way this man speaks." (John 7:46) His command of the Scriptures was spoken as one with **authority** because He is the Scriptures' **author**.

In the book of Acts, we have a clear demonstration of Jesus' teachings in action primarily through the apostles. Author Luke traces the expansion of the Christian movement from its earliest beginnings hallmarked by the new-found boldness of Peter and martyrdom of Stephen to the time it reached worldwide proportions, primarily through the ministry of Luke's companion Paul. The Apostle Paul would go on to compose a series of letters to churches he helped establish on his missionary journeys that ultimately would be canonized into the pages of the New Testament, and he, too, would willingly go to his death for his convictions.

Finally, if Jesus was easily dismissed as a lunatic, doesn't it seem incredibly unlikely that for centuries thereafter His teachings would spawn countless books, poems, and songs, inspire and transform billions of people, spur the foundation for myriads of outreach ministries, soup kitchens, hospitals, orphanages, etc.?

It's worth making the point that some people who lived during the time in which He lived did think Him insane (John 10:20), including, at first, His own brother James (Mark 3:21). By the way, in addition to being accused of being deranged, He was also accused of being deceitful (John 7:12), drunk (Matthew 11:19), and demon-possessed (Mark 3:22) at various times by both his family and his enemies. I find it encouraging, refreshing, and brutally honest that the writers of the gospels and other parts of the New Testament didn't shy away from this possibility and, in fact, included it in their writings for generations of readers to see. As for brother James? He became a radical follower of Christ and prominent leader in the early church after Jesus appeared to him in His resurrected state.

How plausible is it that He was a liar? First and foremost, would a liar allow himself to be ridiculed, spit upon, beaten, bludgeoned, scourged, and crucified just to promote a deception? You'd **really** have to be a nut. Also, liars usually crack under pressure and change their story, especially when their lie no longer serves their self-interests. They virtually always have motive for lying that is self-serving, unless they are simply pathological liars, which would imply a degree of mental instability.

In Jesus' case, rather than having selfish ambitions, His outrageous claims brought to Him no personal benefit — only negative consequences. Whereas cult leaders often seek power and control, Jesus declined both during His ministry. He was not motivated by money; he was poor and stood up for those in His society who were poor and marginalized. (16)

Another point worth mentioning is that liars typically don't display such remarkable moral virtue as Jesus did. He spoke the truth and encouraged others to do the same. Philosopher Douglas Groothius writes, "Could Jesus be both a great (many would say the greatest) moralist as well as the greatest liar of all time? The question answers itself." (17)

By the way, why is it, do you think, that the name of Jesus evokes so much guttural indignation, division, and hostility among non-believers? If He was just an ideologue, lunatic or liar, why does His name elicit so much emotion? It's not as if He was an evil man, such as Hitler, to justify those feelings. To me, that reaction alone is a sign that His Lordship is authentic, because it reinforces our natural desire to reject His authority in our lives. Why else should someone get so heated? It is something worth thinking about. Do people get so upset when someone mentions the name of David Koresh or Moon Sun Myung? I suspect many people don't even know who these people were? They both claimed to be God. We rarely talk about them and certainly don't sing songs about them. Because their conduct and motives ultimately revealed there was no truth to their claims, they have been easily dismissed. Jesus' claims, however, have been validated, thus forcing every man to confront their implications.

Before moving on to the last option, namely, that Jesus is Lord, some might be thinking (and I hear this a lot), "Wait just a minute! I believe He was just a good moral teacher!" Well, that sounds good but is not an option, and here's why. He claimed to be God... Author of life, Creator of all things with power over sin and death, the Beginning and the End, and the only Way to the Father. These claims are either true or they aren't. If true, then He is far more than a good moral teacher. He is God! If they are not true, then He is a monumental liar at best, which undermines any credibility as a reliable moral teacher.

Let's consider an uncommon fourth option proposed by some skeptics; that is, that Jesus is not Lord, liar, or lunatic, but instead a *legend* manufactured by exaggerations of the early church. This option seems preposterous for a number of reasons, some of which I will simply list.

- There are more biographies written of Jesus than anyone from the ancient world. (18)

- The gospels are replete with specific historical details that have been confirmed by the modern science of archaeology.

World renowned scholar Gleason Archer authoritatively states, "As I have dealt with one apparent discrepancy after another and have studied the alleged contradictions between the biblical record and the evidence of linguistics, archaeology, or science, my confidence in the trustworthiness of Scripture has been repeatedly verified and strengthened by the discovery that almost every problem in Scripture that has ever been discovered by man, from ancient times until now, has been dealt with in a completely satisfactory manner by the biblical text itself—or else by objective archaeological information." (19)

- Even non-Christian historians, including the famous first century Roman historian Josephus, believed that Jesus claimed to be God and the Jewish Messiah. (20)

- In writing the gospels, the disciples did not paint a glamorous picture of themselves (referring to themselves as unintelligent (Mark 9:32, Luke 18:34), uneducated (Acts 4:13), uncaring (Mark 14:32), cowardly (Matthew 26:33-25), and doubtful (Matthew 28:17). Peter even mentions that Jesus referred to him as Satan himself (Mark 8:33). The disciples also included the fact that women were the first eyewitnesses of the resurrection in a day when women were second class citizens—unable to testify in a court of law. Josephus writes, "But let not the testimony of women be admitted, on account of the levity and boldness of their sex, nor let servants be admitted to give testimony on account of the ignobility of their soul; since it is probable that they may not speak truth, either out of hope of gain, or fear of punishment." (21) The apparent transparency in these accounts serves to actually enhance their credibility, and are unlikely to appear in a serious attempt at fabricating a believable story.

- Inventing a divine Jesus meant a certain savage death to the disciples and others who propagated the gospel message if caught by the Roman authorities since the law required the death penalty to anyone who denied the self-proclaimed divinity of the Roman Emperor. In fact, Christians were the common objects of sport in the gruesome games of the Colosseum, and, in the days of Emperor Nero, were impaled, covered with pitch and hot wax, and used as human torches to light the path to his palace among other equally ghastly atrocities. (22)

Now let's presume Jesus *is* God in the flesh and the Messiah*, the Chosen One sent to earth to redeem people of their sins. Once again, I turn to fulfilled prophecy to support the case. In 2 Peter 1:18-19, the Apostle Peter writes, "...and we ourselves heard this utterance made from Heaven when we were with Him on the holy mountain. So we have *the prophetic word made more sure* [emphasis mine], to which you do well to pay attention as to a lamp shining in a dark place, until the day dawns and the morning star arises in your hearts."

* Christians believe that Jesus is the Jewish Messiah. Messiah (literally, "Anointed One") and translated Kristós in Greek (from which we get the term "Christ" in English), is the Son of God; that is, a visible manifestation of the invisible God and, indeed, One and the same as God the Father (Hebrews 1:1-3). Hence, the phrases "Jesus is Messiah" and "Jesus is God" are used interchangeably in this narrative. Moreover, prophecies fulfilled by Jesus the Messiah are, too, visible manifestations of the invisible God and affirming proofs of His claims.

What did Peter say? Here is a guy who was a disciple of Jesus for more than 3 years. He walked with Him, talked with Him, ate with Him, learned from Him, and was in His inner circle, experiencing some events that only he and two others (James and John) were privy to. And it is he who says, "…the prophetic word is **more sure.**" Why would prophecy be important to consider compared to actually being with the Messiah in person? Let's see why.

Using composite probability—simply stated, the mathematical union of two or more independent events—let's examine the probabilities supporting that Jesus is the Messiah.

First, let me employ a commonly used example to illustrate composite probability. Let's pretend there is a room containing 100 people composed of 40 men and 60 women. The odds that a person who walks out the door is a man is 40/100 or 40%. Likewise, the chances that it is a woman is 60/100 or 60%. Let's also pretend that 20 people in the room are left-handed and 80 are right-handed. The odds that a left-handed person exits the room is 20/100 or 20%. Likewise, the odds of a right-handed person leaving is 80/100 or 80%.

What are the odds that a left-handed woman exits? This calculation employs the principle of composite probability. The odds become 60% (or 0.6) that the person is a woman multiplied by 20% (or 0.2) that the person is left-handed. The odds become 0.6 x 0.2 = 0.12 or 12%.

Back to Jesus.

In short, the Hebrew Scriptures (Old Testament) contain **more than 300 prophecies detailing the coming Messiah...100% of which were fulfilled by Jesus Christ**. Isn't that remarkable? More than 300! "So what makes that such a big deal?" one might ask. Well, let's take a look at just 8 of these prophecies:

- **The Messiah will be born in Bethlehem**

  o Old Testament Prophecy: *Micah 5:2: "But you, O Bethlehem Ephrathah, are only a small village in Judah. Yet a ruler of Israel will come from you, one whose origins are from the distant past."*

  o New Testament Fulfillment: Matthew 2:1, Luke 2:4-7

- **The Messiah will be born of a Virgin**

  o Old Testament Prophecy: *Isaiah 7:14: "Therefore the Lord Himself will give you a sign: Behold, a virgin will be with child and bear a son, and she will call His name Immanuel."*

  o New Testament Fulfillment: Matthew 1:24-25

- **The Messiah will be a prophet like Moses**

o   Old Testament Prophecy: *Deuteronomy 18:15: "The Lord your God will raise up for you a prophet like me [Moses] from among your fellow Israelites, and you must listen to that prophet."*

o   New Testament Fulfillment: John 1:45

- **The Messiah will enter Jerusalem triumphantly on a donkey's colt**

o   Old Testament Prophecy: *Zechariah 9:9: "Rejoice greatly, O people of Zion! Shout in triumph, O people of Jerusalem! Look, your king is coming to you. He is righteous and victorious, yet he is humble, riding on a donkey — even on a donkey's colt."*

o   New Testament Fulfillment: Matthew 21:1-11

- **The Messiah will be rejected by His own people**

o   Old Testament Prophecy: *Isaiah 53:1, 3: "Who has believed our message? To whom will the Lord reveal His saving power? He was despised and rejected – a man of sorrows, acquainted with bitterest grief. We turned our backs on Him and looked the other way when He went by. He was despised, and we did not care."*

o   New Testament Fulfillment: Luke 23:20-24

- **The Messiah will be betrayed by one of His followers**

o   Old Testament Prophecy: *Psalm 41:9: "Even my close friend, whom I trusted, he who shared my bread, has lifted up his heel against me." (NIV)*

o   Old Testament Prophecy: *Psalm 55:12-13: "It is not an enemy who taunts me – I could bear that. It is not my foes who so arrogantly insult me – I could have hidden from them. Instead, it is you — my equal, my companion and close friend."*

o   New Testament Fulfillment: John 18:4-5

- **The Messiah will be betrayed for 30 pieces of silver**

o   Old Testament Prophecy: *Zechariah 11:12-13: "And I said unto them, 'If ye think good, give me my price; and if not, forbear. So they weighed for my price thirty pieces of silver. And the Lord said unto me, Cast it unto the potter: a goodly price that I was prised at of them. And I*

> *took the thirty pieces of silver, and cast them to the potter in the house of the Lord." (KJV)*

o New Testament Fulfillment: Matthew 26: 14-16

- **The Messiah will sojourn from Egypt**

o Old Testament Prophecy: *Hosea 11:1: "When Israel was a child, I loved Him, and out of Egypt I called my Son."*

o New Testament Fulfillment: Matthew 2:13-15, John 19:14-15

What are the odds that one man could fulfill all 8 of these prophecies? Let's apply ***very conservative probabilities*** to each of these prophecies. For example, regarding his birthplace prophesied by the prophet Micah: Based on a prophecy that the Messiah would be born at a specific time in history (which we will address shortly), what is the probability that Jesus would be born in Bethlehem during that time?

For ease of calculation, we can assume that the average birth rate in Bethlehem in that period of time is proportionate to its average population. The population of Bethlehem in Jesus' day was generously about 1000 people (more likely around 200). Indeed, it was "the little town of Bethlehem." Estimated world population at the time of Jesus' birth from a variety of sources ranges from 150-200 million. (23) We could be ultra-conservative and say that the world's population was only 100 million. Therefore, the odds of anyone being born in Bethlehem, as opposed to anywhere else in the world, at that time is approximately 1,000/100,000,000 or $1:10^5$.

As if these odds were not remarkable enough, taking into account that Jesus' parents, Mary and Joseph, were not living in Bethlehem at the time (but rather approximately 75 miles north in Nazareth), makes the likelihood of Jesus being born in Bethlehem, and thus fulfilling this prophecy, even more astounding.

For the sake of space and time, if we apply similar conservative odds to each of the next 7 prophecies and divide that number by the estimated total population of the world since the beginning of time (approximately 100,000,000,000 or 10^{11}), the odds of one person (Jesus) fulfilling all 8 prophecies is *conservatively* estimated at $10^{28}/10^{11} = 1:10^{17}$.

What exactly does that mean? Glad you asked. I'll use a simple, common analogy that helps illustrate. Let's say you were going to reach in and attempt to pull the one red-colored coin from a bucket of otherwise-silver coins.

To convey these odds ($1{:}10^{17}$), how large would your bucket have to be to hold that many silver coins? Someone, blessed be he, took the time to calculate that such a bucket, assuming it was two feet deep, would have to be the size of the state of Texas. In case that hasn't sunk in...THAT'S HUGE!

Imagine roaming around the state of Texas for a while in a pile of silver dollars 2 feet deep, blindfolded, and, when you feel like it, sticking your hand down and picking up a coin that happens to be the red one. (Figure 1) The odds of that happening is the same. And that's just for 8 prophecies!

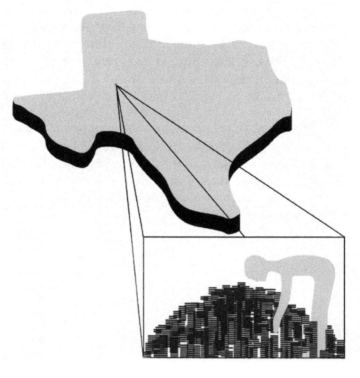

Figure 1

How about 8 more prophecies?

- **The Messiah will be tried and condemned to death**

40

- o Old Testament Prophecy: *Isaiah 53:8: "From prison and trial they led Him away to His death. But who among the people realized that He was dying for their sins — that He was suffering their punishment?"*

- o New Testament Fulfillment: Luke 22:66-71, 23:13-25

- **The Messiah will be silent before His accusers**

- o Old Testament Prophecy: *Isaiah 53:7-8a: "He was oppressed and treated harshly, yet He never said a word. He was led as a lamb to the slaughter. And as a sheep is silent before the shearers, He did not open His mouth. From prison and trial they led Him away to His death."*

- o New Testament Fulfillment: Matthew 26:62-63

- **The Messiah will be smitten and spat upon**

- o Old Testament Prophecy: *Micah 5:1: "Mobilize! Marshal your troops! The enemy is laying siege to Jerusalem. With a rod they will strike the leader of Israel in the face."*

- o Old Testament Prophecy: *Isaiah 50:6: "I give My back to those who beat Me and My cheeks to* those who pull out My beard. I do not hide from shame, for they mock Me and spit in My face."

o   New Testament Fulfillment: Mark 14:65, 15:19

- **The Messiah will be mocked and taunted**

o   Old Testament Prophecy: *Psalm 22:7-8: "Everyone who sees Me, mocks Me. They sneer and shake their heads, saying, 'Is this the One who relies on the Lord? Then let the Lord save Him! If the Lord loves Him so much, then let the Lord rescue Him!'"*

o   New Testament Fulfillment: Matthew 27:29-31.

- **The Messiah will die by crucifixion and be pierced**

o   Old Testament Prophecy: *Psalm 22:14-16: "My life is poured out like water, and all My bones are out of joint. My heart is like wax, melting within Me. My strength has dried up like sun baked clay. My tongue sticks to the roof of My mouth. You have laid Me in the dust and left Me for dead. My enemies surround Me like a pack of dogs; an evil gang closes in on Me. They have pierced My hands and feet."*

o   Old Testament Prophecy: *Zechariah 12:10a: "Then I will pour out a spirit of grace and pray-*

er on the family of David and on all the people of Jerusalem. They will look on Me whom they have pierced and mourn for Him as for an only son." *

o New Testament Fulfillment: Luke 24:29-40, Mark 15:25, John 19:34.

• **The Messiah's garments will be divided by casting lots**

o Old Testament Prophecy: *Psalm 22:18: "They divide My clothes among themselves and throw dice for My garments."*

o New Testament Fulfillment: Matthew 27:35, Mark 15:24

• **The Messiah will suffer with sinners**

o Old Testament Prophecy: *Isaiah 53:12a: "I will give Him the honors of One who is mighty and great, because He exposed Himself to death. He was counted among those who were sinners."*

o New Testament Fulfillment: Matthew 27:38, Mark 15:27

* By the way, this particular prophecy was written 700 years before crucifixion was even invented!

- **The Messiah's bones will not be broken**

o Old Testament Prophecy: *Numbers 9:12: "They must not leave any of the lamb until the next morning, and they must not break any of its bones. They must follow all the normal regulations concerning the Passover."*

o *Psalm 34:20: "He keeps all his bones, not one of them is broken."*

o New Testament Fulfillment: John 19:31-33

Applying the same principles, the odds of all 16 prophecies being fulfilled in one person becomes 10^{28} x $10^{28}/10^{11} = 10^{45}$ Guess how big the bucket is now? How about ***30 times the distance of the earth to the sun, which turns out to be about 2,790,000,000 miles!*** (24) This is roughly equivalent to the distance between the sun and Neptune. (Figure 2)

There are actually more than 300 prophecies fulfilled in Jesus Christ; many dozens of more obscure prophecies that make the odds even **more** staggering!

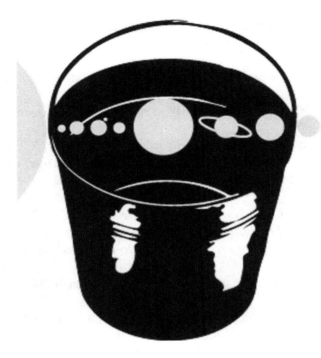

Figure 2

How about a couple of more prophecies for fun?

- **The Messiah will be buried in a rich man's tomb**

o   Old Testament Prophecy: *Isaiah 53:9: "He had done no wrong, and He never deceived anyone.*

But He was buried like a criminal; He was put in a rich man's grave."

o   New Testament Fulfillment: Matthew 27:57-60

- **The Messiah will be raised from the dead**

o   Old Testament Prophecy: *Psalm 16:10: "For You will not leave my soul among the dead or allow Your Godly One to rot in the grave."*

o   Old Testament Prophecy: *Psalm 30:3: "You brought me up from the grave, O Lord. You kept me from falling into the pit of death."*

o   New Testament Fulfillment: Acts 10:39-41

Raised from the dead! That bears repeating...***raised from the dead...***since the resurrection is the hinge on which the door of Christianity swings. For this reason, the resurrection of Jesus has probably been studied and scrutinized more than any other prophecy related to His life as skeptics attempt to disparage the Christian faith. To this day, its validity has not been disproven. For a very comprehensive discussion of this topic, refer to *The Resurrection of Jesus: A New Historiographical Approach by Mike Licona referenced in Appendix A.*

Here are a couple of other points to further strengthen the argument of Jesus' authenticity as the risen Messiah:

- The book of 1 Corinthians was penned by the Apostle Paul somewhere between 50-60 AD according to most historians, before any of the gospels were written. In chapter 15 regarding Jesus' resurrection, Paul writes, "For I delivered to you as of first importance what I also received, that Christ died for our sins according to the Scriptures, and that He was buried, and that He was raised on the third day according to the Scriptures, and that He appeared to Cephas [Peter], then to the twelve. After that He appeared to *more than five hundred brethren* [emphasis mine] at one time, most of whom remain until now, but some have fallen asleep; then He appeared to James, then to all the apostles; and last of all, as to one untimely born, He appeared to me also." (1 Corinthians 15:3-8)

  Jesus was crucified on the cross sometime between 30-35 AD. Less than 30 years later, Paul recorded the words above while, undoubtedly, some subset of the people (perhaps many) to whom Jesus appeared in His resurrected state were still alive. That's bold, since one testimony to the contrary might have cast a serious shadow on the validity of his writings for all time and might have been the catalyst to debunk Christianity altogether. Yet not one did.

- There are recorded in the pages of Scripture amazingly detailed genealogies of Jesus through the line of King David and dating back to Adam that could have easily been challenged, especially in a culture where scribes kept meticulous transcripts and historical archives.

How about this mind-boggling prophecy? Written before 500 BC, the prophet Daniel writes, "So you are to know and discern that from the issuing of a decree to restore and rebuild Jerusalem until Messiah the Prince there will be seven weeks and sixty-two weeks; it will be built again, with plaza and moat, even in times of distress. Then after the sixty-two weeks the Messiah will be cut off and have nothing, and the people of the prince who is to come will destroy the city and the sanctuary…" (Daniel 9:25-26)

The idiom "weeks" are not literal weeks, but, rather, in Hebrew prophetic parlance, the word represents a period of time multiplied by seven. It's similar to the phrase "69 dozen" representing a multiple of twelve or "69 tons" representing a multiple of 2000. In this case "weeks" refers to a multiple of seven years. So, Daniel prophesied that the Jewish Messiah would present Himself as King exactly 483 years (7 + 62 = 69 weeks = 69 years x 7 = 483 years) after the issuing of a decree to restore and rebuild Jerusalem.

There is abundant documentation to demonstrate that the official order to restore and rebuild Jerusalem was issued by Artaxerxes Longimanus on March 14, 445 BC. [Emphasis on the "plaza and moat" distinguishes this decree from other earlier mandates to rebuild the Temple.]* (25) Accounting for leap years, transition from 1 BC to 1 AD (only one year since there is no zero year), and a 360-day calendar year used by the Jews and Babylonians, the day predicted by Daniel's prophecy for the Christ to be acknowledged as King becomes April 6, 32 AD, the exact day noted by multiple ancient historians of Jesus' "triumphal entry." This is the day Jesus rode into Jerusalem on a donkey's colt, prophesied centuries earlier by the prophet Zechariah (Zechariah 9:9), to the shouts of the crowd's praises and waving of palm branches (Luke 19:28-44); and the day millions of people around the globe today celebrate as Palm Sunday.

Furthermore, the prophecy states that the Messiah would be "cut off" before the sacking of the Temple. I had to read that one carefully. *According to biblical*

* At least 3 decrees were issued. Most biblical scholars agree that this is the correct one.

prophecy, **the Messiah had to die** *(and violently at that, since "cut off" in the Hebrew text implies a violent death)* **before the destruction of the Temple in Jerusalem, which took place in 70 AD.** *That* certainly narrows the field of messianic candidates!

Taking just 48 prophecies and applying the math, the odds now become 10^{157}.

As Chuck Missler of Koinonia House describes it, the "bucket" now is composed of a ball comprised **OF** every atom in the universe (which, as noted in the last chapter, by scientist's calculation, is estimated at 10^{66}) **FOR** every atom in the universe (10^{66}) repeated for every second since the universe began (for evolutionists, estimated at 15 billion years or about 10^{17} seconds). (26) *Please read that sentence again to grasp the enormity of the numbers we're discussing.* That equals 10^{66} (OF every atom) x 10^{66} (FOR every atom) x 10^{17} (TOTAL seconds) = 10^{149} (See Figure 3) (I never imagined my math degree would come in handy). And we're still short of 10^{157} by 100,000,000 times!

Considering that the odds of winning an average state lottery are generally 10^8 (often more), the odds of Jesus fulfilling just these 48 prophecies is roughly equivalent to winning the lottery 20 days in a row.

OF
EVERY ATOM IN
THE UNIVERSE

FOR
EVERY ATOM IN
THE UNIVERSE

FOR
EVERY SECOND
SINCE THE WORLD
BEGAN

$$\times \frac{}{10^{149}}$$

Figure 3

If you believe in a young earth like I do (about 6,000 years old), then the number of seconds that have ticked off since it was created is less, roughly 10^{11}, so that the bucket is then only 10^{143} ($10^{66} \times 10^{66} \times 10^{11}$) — still quite an astronomical figure. Those are the odds for only 48 prophecies fulfilled by one person. I won't bore you to calculate 96 or 192, (forget 300)! I think you get the point. Our bucket of coins is now thousands of miles deep and spans the entire universe which is billions of light years across. These are pretty convincing odds; don't you think? Honestly, with these odds, how hard is it to believe that Jesus is who He said He was and is?

Yet, many are still persuaded into dismissing these odds and disbelieve. I suggest that such persuasion is Satan's strategy (as the Bible predicts). There are multiple verses in the Bible that describe Satan as a deceiver and liar, perhaps the most explicit of which is John 8:44 which reads, "...Whenever he speaks a lie, he speaks from his own nature, for he is a liar and the father of lies."

There is nothing new under the sun. Since the onset of human history, Satan has been deceiving people, questioning God's authority, and casting doubt on what God plainly tells us, daring us to actually believe it. In Matthew 4:3, the devil challenges Jesus saying, "If you are the son of God...," an objection that comes on the heels of God proclaiming audibly that Jesus was His "dearly beloved Son."

Among his most notorious lies are the following:

- There is no God.
- There is no hell.

- God grades on the curve.
- Good deeds get you into Heaven.
- Many roads lead to Heaven.
- A little sin doesn't hurt anyone.
- Happiness can be achieved by what we acquire.
- Truth is relative.
- God doesn't love you.
- There is no hope for you in this world.

So, what will you believe? The indisputable numbers or Satan's lies?

One final point: As incredible as these odds are, there are **even more** Scripture verses prophesying Jesus' Second Coming (His return to earth as King) than there are His first. Once again, from *What is the Truth? The Case for Biblical Integrity (Kindle Edition),* Dr. Chuck Missler points out that there are, in fact, over 1800 references in the Old Testament to Jesus' Second Coming and 318 references found in 23 of 27 books that comprise the New Testament. In other words, for every prophecy of His first coming in the Old Testament, there are 8 predicting His Second Coming. (27)

If one remains a non-believer in the face of all that has been presented here, what possible excuse could they offer Christ when He does return, because the odds are He will!

"Anyone who isn't with Me opposes Me,

and anyone who isn't working with Me

is actually working against Me."

-Matthew 12:30
(New Living Translation)

Chapter 6: Taking it Home

Why did I take the time to write this book? To quote Chris Tiegreen from his devotional entitled *God With Us,* 'For those of us who believe, it soon becomes painfully alarming to us that we know plenty of people who do not. We who count on the promises of God cannot be content to keep them to ourselves. That would be like dining at a never-ending, everyone's - invited buffet, but keeping it a secret from starving people lingering around us." (28) More succinctly, it's not possible for Christians to know what we know and remain unconcerned about those who don't.

Hopefully, by now, I have demonstrated that the step of faith to believe that the Bible is reliable and that Jesus is Lord is actually nothing more difficult than a tiny wiggle of your toe. How easy does it become, then, to do what Acts 16:31 says, "Believe in the Lord Jesus Christ and you will be saved…" Just believe. Based on what has been presented, I hope you'll agree that it takes **more** faith **not** to believe that Jesus is Lord.

Apply the odds from previous chapters to anything else in life — investments, business decisions, weather, horse races, lottery tickets — and anyone would be sold! But apply it to eternal destiny and those same people balk. ***Now that's odd!***

Too often it seems this hesitation to "buy-in" is simply a choice to avoid the issue. While this life is but a blip on the timeline of eternity, I see people spend nearly every waking moment concerned about *it*, rather than the life to come. I have asked myself, "Why spend lots of time accumulating and worrying about things that won't last?" — which will pass away like a vapor (James 4:14). Jim Elliot, the great missionary to Ecuador in the 1950s, expressed it like this, "He is no fool who gives what he cannot keep to gain that which he cannot lose." I love that. And John Tillotson summed it up when he said, "He who provides for this life, but takes no care for eternity, is wise for a moment, but a fool forever." (29)

In addition to the good things in life to which we cling, the difficulties we face must also be kept in perspective. The Bible says that the tribulations of life are light and momentary compared to the glory that is to be revealed (2 Corinthians 4:17-18). The challenges we face in this life will pass, but the worthiness and relevance of Jesus Christ will endure forever. Eternal life is not some fantasy reserved for those who have discovered the proverbial fountain of youth. It is not reserved for the "good or righteous," but available freely to all…. right now. Just believe. Isn't it something worth considering?

In Revelation 3:20 Jesus says, "Behold, I stand at the door and knock; if anyone hears My voice and opens the door, I will come in to him and will dine with him, and he with Me." I believe Jesus is knocking on the door of your heart. Can you feel it? Do you want to believe? If so, then please get on your knees and tell Him that you believe the Bible is true and that He is Lord. Ask for His forgiveness for not believing until now. Open the door of your heart and welcome Him in. That's all it takes.

I made that decision on November 11, 1989, at the age of 29, and have never regretted a moment. He took a selfish, skeptical, prideful, foul-mouthed, narcissistic liar who was dabbling in all sorts of unhealthy activities and washed me clean.* In the book of 1 John, verse 1:9 says, "If we confess our sins, He is faithful and righteous to forgive us our sins and to cleanse us from all unrighteousness." ***That's good news!*** It was for me.

Please make the choice now. The same Bible that predicted that the Messiah would come in the person of Jesus Christ back in the first century says that He is coming again...soon. The second time He is coming to welcome those who believe into His Kingdom, but also to administer judgment to those who have rejected the truth of who He is.

* For my full testimony, see Chapter 8.

You can count on it! The numbers don't lie and, as you have already seen, it's a losing strategy to bet against the odds of biblical prophecy coming true. So be among the believers and take hold of eternal life.

The Bible makes it clear that every person who has ever drawn a breath must face the sovereign and just God. It matters not whether you are man or woman, young or old, rich or poor, famous or common, educated or illiterate, powerful or ineffective, ambitious or idle. All will be held accountable for their life, and anything short of perfection will merit His judgment.

Hebrews 9:27 (New International Version) says, "Just as people are destined to die once, and after that to face judgment..." Since none of us is perfect, we must rely on Jesus' perfect sacrifice on our behalf to stand before God with confidence.

Quoting from online notes administered by www.bsfinternational.org, in Revelation Lesson 26, p. 5:

"John's revelation says eternity is real; there is a heaven to be gained and a hell to be avoided through faith in the Son of God. Popular opinion resists the idea that God will judge those who reject Jesus Christ. But to reject the reality of judgment demeans the holy God and lessens the value of human life. God has revealed Himself personally in His creation, His Word and His Son.

His truth demands a response. Because God created people in His own image*, no human life is insignificant. All that people think and do matter to God. He cannot ignore wickedness, rebellion or apathetic rejection of all that is good. How could God be just and not judge those who despise His Son?"

In light of this, nothing is more important than your response to the gospel message. You are responsible without excuse for ignored opportunities. Jesus' words and claims, laden with power and purpose, require a decision. They cannot be treated casually. Please hear me when I say that it is the most disastrous mistake of your life to reject the Savior Jesus, and it is wishful thinking to believe that you can be saved based on your own accomplishments and achievements apart from faith in Him.

The consequences for rejecting God's free gift of salvation through Jesus Christ is an eternity apart from Him. This is hell. Hell was originally created for rebellious angels (2 Peter 2:4), and it was never God's desire that it be for us. But that is our destiny if we choose not to believe (Matthew 25:41) and spurn His authority. Bear in mind that *even the demons believe that Jesus is the Son of God* (Mark 1:24, 34, James 2:19), but refuse to bow to His Lordship and are destined to eternal punishment.

* Genesis 1:27

The fact that God has not *already* judged humanity as detailed in the book of Revelation is what amazes me most. We are deserving of His wrath, but He has postponed His judgment thus far because He desires that **none** should perish (2 Peter 3:9). He is giving everyone every opportunity to accept His free gift of salvation. Perhaps this book is your opportunity. Don't let another day go by without being assured of your eternal salvation.

So, **believe** what God has promised, **trust** Him, and **rest** in His promises. Take a few minutes someday and explore the promises of God. They are abundant, but include eternal life, forgiveness, peace, provision, guidance, wisdom, protection, and deliverance. These things are available to you for the asking. Don't complicate things. It's simple. Just believe. *The odds are in your favor*!

Chapter 7: Final Word

Some who are reading this book may be atheists or agnostics. So was I. If you are in that category I say, "Thank you for reading this book." Regardless of what anyone presently believes or doesn't believe, we are all equally precious to God and the gospel message is pertinent and applicable to all of us. Don't believe the lie any longer. *Jesus is the only way* to the Father, and *all* can come to the Father through Him (John 14:6, Acts 4:12). In fact, God desires nothing more — so much so, that He sacrificed His only Son for you.

Please realize that God did not intend for humanity to die. We chose our path and the results are all around us. Pick up the newspaper and you can't help but be discouraged by the brokenness of this world. Separation from God, due to sin, is the most pressing problem humanity faces, and there is no human solution (30). But God has provided a way to have a restored fellowship with Him.

Some may not believe in or trust God anymore because of circumstances in their lives, perhaps a tragedy that they've had to endure. Such feelings are understandable. My answer to that is, "God is still on the throne and He has a perfect plan for your life." Try not to let a perceived failure in His trustworthiness cloud

61

your view. He is worthy of our trust. What we consider adversity is sometimes God's means of drawing us into a closer relationship with Him. Remain hopeful and trust in His unwavering, eternal promises (Hebrews 10:23).

Often, between the spoken promises and their fulfillment rages the battle of faith, and a prime weapon of our adversary, the devil, is doubt. Don't doubt God. Numbers 23:19 says, "God is not a man, that He should lie, nor a son of man, that He should repent; Has He said, and will He not do it? Or has He spoken, and will He not make it good?" God's promises are the reality behind the appearance, the real truth behind the veneer of this life. They are more real than anything you or I see or anything our other senses perceive; more real than anything our minds can conceive. If we are called to follow an invisible God, then we must walk by faith and not by sight (Hebrews 11:1).

God's purpose in history is to reveal His glory. He has done it in creation. Consider its marvelous, magnificent, and perfect order. In fact, creation speaks so clearly of God's presence and power that the Bible says, for that reason alone, people are without excuse before their Maker (Romans 1:20). He has done it in history by His sovereignty over the rise and fall of nations. He has done it in the birth, life, death, and resurrection of His Son and the glorious love that motivated Him. He has done it through the salvation and transformation of people throughout the ages who have believed in Him, and He does it daily via the Holy Spirit through redeemed believers today. Everything exists because God made it; everything continues to exist because God sustains it. He created you and ordained the days of your life (Psalm 139:13-16). You are able to read this page because God has given you sight. Do you realize that we could not take our next breath unless He allowed it?

It's time to get serious. This is not a joke. I don't think I need to remind anyone that life is short. And the years seem to go by faster as we grow older. Whether you're under spiritual conviction or not, your eternal destiny weighs in the balance. Would you unnecessarily risk something as serious as that? The bottom line is this: either you believe that Jesus Christ, as God and Messiah, bore your deserved eternal punishment for sin on the cross on your behalf nearly 2,000 years ago or you will stand condemned and have to bear it yourself. If you believe what you've read in this book and trust that Jesus is Lord, then you will not only escape the punishment and horrors of hell, but experience true and eternal peace with God.

If you think about it, life begins with our first breath and we begin our march toward death starting with the second breath. None of us are promised another breath tomorrow. So, make today your day of salvation! (2 Corinthians 6:2) (see Appendix C for a sample prayer)

I have been to Jerusalem and stood in Jesus' tomb. It's empty. I can assure you. He is the risen Lord. And His victory over death assures your victory, too. Simply believe!

Notes

- All Scripture verses are taken from the New American Standard translation of the Bible unless otherwise specified.

- I welcome comments, criticism, questions, and personal testimony. Please send your remarks to whataretheoddswjv@gmail.com

Chapter 8: Personal Testimony

Growing up, our family attended the Greek Orthodox Church. Despite attending church practically all my life—regularly as a child and adolescent, and more sparingly as a teenager—I don't recall ever clearly and understandably hearing the gospel message. For several years prior to the time that I was ultimately confronted with the truth, I had become dismayed with the habitual, mechanical nature of the church services, and so my attendance at church and faith in God had waned to the point of almost being non-existent when I set off for college.

Throughout college and medical school, I barely gave God a thought. The medical curriculum, emphasizing such things as evolutionary biology, reinforced the unlikelihood of intelligent design, and I found myself drifting toward agnosticism. Upon completion of medical school, I moved from New York to Miami to begin a rigorous residency program.

During my early training, aside from being disillusioned with church itself, my weekends had become increasingly precious to me, and I took advantage of my free time. My church attendance was essentially reduced to twice a year, on Easter and Christmas. I was officially a "two-timer."

In general, things were going well for me. I was enjoying life in a beautiful place. Yet, paradoxically, I felt dissatisfied and hollow, though I could not attribute my feelings to anything specific.

About six months after arriving in Florida I received a phone call from a distant cousin who had been living in Coral Springs, about 45 minutes north of my home. I accepted her invitation to dinner, took a break from the parties, and joined her one Friday night. Little did I realize that the course of my life would change that evening.

After casual conversation and small talk, the topic turned to religion, and she bluntly asked me, "If you die tonight, are you 100% sure you'll go to Heaven?"

"Sure I'm sure," I replied, though deep down I had my doubts.

"What makes you so sure?" she asked. I proceeded to list all of my qualifications for entrance into the Kingdom of God, including being "a nice guy," none of which seemed to make any impression on her whatsoever.

When I was done with my lofty self-assessment, she looked me in the eye with sadness and compassion and a tear in hers and simply said, "I'm sorry, but since

I love you, I have to tell you that you're wrong." I could sense that she was not being antagonistic, but had a deep concern for my welfare and was truly speaking in love. Still, I didn't easily surrender to her analysis and came out swinging.

Needless to say, we talked for several hours and literally closed the restaurant. During our conversation, it became painfully apparent to me that she was speaking with a sense of authority, often quoting Scripture, with which I was unfamiliar but intrigued. I, on the other hand, was simply stating my opinion. I vowed that night to her and to myself that I was going to begin reading the Bible with regularity, if for no other reason, to justify my feelings and lifestyle and prove her wrong.

My best laid plans came to a screeching halt within a few days as I found Bible reading tedious and dull. To her credit, my cousin called me every so often to see how I was progressing and never spoke judgmentally when I gave her my feeble excuses about why my Bible was back on the shelf. The truth is that I didn't think it was that important, because I always found time for the things I considered important. In her last phone call to me, she asked me to pray for someone to come into my life locally to help me understand more clearly the things she tried to convey during our dinner date. I promised that I would, and so I did that very night. It seemed harmless enough and, besides, I didn't expect any results. I don't even recall exactly what I prayed, but it was heartless. Regardless, apparently the Lord heard my prayer because the very next day would be one that I would never forget.

I wasn't even on the schedule to work until late the following evening and my plan was to sleep as late as possible. Yet, at the crack of dawn, my phone rang and woke me abruptly. It was a fellow resident who pleaded with me to work for him because he had unexpectedly developed a tooth abscess the night before that was excruciatingly painful and required immediate attention. My pleas for him to call someone else fell on deaf ears and I found myself working at the hospital that fateful morning.

I was in training as a radiologist and had just completed a procedure on a very sick patient. I was back in the reading room dictating the case when a team of physicians entered my work area seeking the results on the patient I had just seen. After a short discussion, off they went down the hall with the exception of a very attractive young woman who simply stood at the entry way, staring at me as if expecting more information. I stared back, all the while wondering how she had escaped my notice these past months. Unsure what to say, I said nothing.

We continued to stare at each other awkwardly for several moments until finally, upon reading the name on my lab coat, she broke the silence and asked me, "Is your name Greek?" My mind was racing as I squirmed to come up with a charming way to ask her on a date. Ordinarily, in my state of mind, that wouldn't have been difficult for me, but, for some inexplicable reason, I was literally unable to speak. All I could manage to do was sheepishly nod my head in the affirmative, and I was becoming increasingly more embarrassed (and sweaty) as I stood silent for what seemed like an eternity.

In desperation, I cried out to God in my mind and pleaded that, if this was the person I prayed about the night before, He unloose my tongue. "Give me the words to say, Lord" I begged in silence. Immediately, I blurted out perhaps the worst pick-up line in the history of mankind. "How important is God in your life?" I asked, expecting her to break into a full sprint.

Instead, she didn't flinch, but simply said, "He's the most important thing in my life." And with that, off she went down the hall to catch up to her team. It was in that moment that the Lord miraculously, but clearly, gave me a vision that I was going to marry this woman. It was not wishful thinking on my part or some starry-eyed daydream. This was a bona-fide, inexplicable and supernatural vision of specific future events, including dates, right up until our wedding day. I was dumb-founded, but convinced that this vision was from God, and told everyone who would listen that I had met my future wife (though I didn't even know her name).

Two days after our first encounter, my wife-to-be found me in the hospital, introduced herself as Jill, and presented me with two gifts that she claimed the Lord told her to bring to me. "The Lord spoke to you?" I asked, not sure I wanted to hear the answer.

"Not audibly" she said, "but I knew in my heart that He wanted you to have these things." "These things" were a more modern version of the Bible that she said would be easier and more enjoyable to read—how did she know I was struggling with reading my Bible? — and a plate of my favorite cookies —how did she know that? Of course, I accepted the gifts, because, for me, it was love at first sight. She then informed me that she was embarking on a 2 to 3-month excursion throughout Europe to visit family and to celebrate, having completed her medical training as a physician assistant, before beginning her first job.

While she was away, admittedly, my motive for reading the Bible was selfish, hoping to win her affection. As time went on, however, the Lord changed my heart and instilled in me a true yearning to know the truth of His Word and to understand it. Much of what my cousin explained to me that night in the restaurant was coming to life and taking root in my heart.

When Jill returned, we met for coffee and I accepted an invitation to attend her church, where they actually taught the Bible verse by verse. Growing up, I don't recall ever even *bringing* a Bible to church, no less studying one. For the first time in my life, the Bible was making sense, and I studied it continually. I became painstakingly aware and convicted of my sin, and it was clear to me that my salvation depended not on my ability to earn God's favor (as I had grown up thinking), but rested solely on the atoning death and resurrection of the Lord Jesus Christ. Subsequently, about 7 months after first meeting Jill, I gave my life to Christ during an altar call at a Christian concert.

Since that time, God has blessed me more abundantly than I could have ever imagined. Jill and I have been happily married now for nearly 25 years and I have two beautiful daughters who are both believers in Christ. God has given me several opportunities to share my testimony and the good news of the gospel, and, as a result, several friends, family members and perfect strangers have been saved.

Now I can say with certainty that I am 100% sure that if I die tonight I will go to Heaven. Can you?

71

Colossians 1:21-22

And although you were formerly alienated and hostile in mind, engaged in evil deeds, yet He has now reconciled you in His fleshly body through death, in order to present you before Him holy and blameless and beyond reproach—

He did this for you

Appendix A: Recommended Readings

Apologetics

- Lewis, C. S. *Mere Christianity*. London: n.p., 1952. Print.

- McDowell, Josh. *More than a Carpenter*. S.l.: Living, 1973. Print.

- Zacharias, Ravi K. *Jesus among Other Gods: The Absolute Claims of the Christian Message*. Nashville, TN: Word Pub., 2000. Print.

- Geisler, Norman L. *Christian Apologetics*. Grand Rapids: Baker Book House, 1976. Print.

- Strobel, Lee. *The Case for Christ: A Journalist's Personal Investigation of the Evidence for Jesus*. Grand Rapids, MI: Zondervan, 1998. Print.

- Geisler, Norman L., and Frank Turek. *I Don't Have Enough Faith to Be an Atheist*. Wheaton, IL: Crossway, 2004. Print.

- Strobel, Lee. *The Case for Christ; The Case for Faith*. Grand Rapids, MI: Zondervan, 2006. Print.

- Rhodes, Ron. *Answering the Objections of Atheists, Agnostics, & Skeptics*. Eugene, Or.: Harvest House, 2006. Print.

- Powell, Doug. *Holman QuickSource Guide to Christian Apologetics*. Nashville, TN: Holman Reference, 2006. Print.

- LaHaye, Tim F., and Tim F. LaHaye. *Why Believe in Jesus?* Eugene, Or.: Harvest House, 2004. Print.

- McDowell, Josh, Josh McDowell, and C. S. Lewis. *The New Evidence That Demands a Verdict*. Nashville, TN: T. Nelson, 1999. Print.

- Groothuis, Douglas R. *Christian Apologetics: A Comprehensive Case for Biblical Faith*. Downers Grove, IL: IVP Academic, 2011. Print.

- Keller, Timothy. *The Reason for God: Belief in an Age of Skepticism*. New York: Dutton, 2008. Print.

- Craig, William Lane., and William Lane. Craig. *Reasonable Faith: Christian Truth and Apologetics*. Wheaton, IL: Crossway, 1994. Print.

- Sproul, R. C. *Defending Your Faith: An Introduction to Apologetics*. Wheaton, IL: Crossway, 2003. Print.

- Licona, Mike. *The Resurrection of Jesus: A New Historiographical Approach*. Downers Grove, IL: IVP Academic, 2010. Print.

- Anderson, Clive, and Brian Edwards. *Evidence for the Bible.* Leominster: Day One Publications, 2014. Print.

Other Recommendations

- Cheney, Johnston M., Stanley A. Ellisen, and Johnston M. Cheney. *Jesus Christ, the Greatest Life: A Unique Blending of the Four Gospels*. Eugene, OR: Paradise Pub., 1999. Print.

- Platt, David. *Counter Culture: A Compassionate Call to Counter Culture in a World of Poverty, Same-sex Marriage, Racism, Sex Slavery, Immigration, Abortion, Persecution, Orphans, and Pornography*. N.p.: n.p., n.d. Print.

- Platt, David. *Follow Me: A Call to Die. a Call to Live*. Carol Stream, IL: Tyndale House, 2013. Print.

- Missler, Chuck W., and William P. Welty. *I, Jesus: An Autobiography*. Coeur D'Alene, ID: Koinonia Institute, 2014. Print.

Appendix B: Relevant Bible Verses

Bible Verses That Demonstrate Jesus' Divinity

Isaiah 7:14: "Therefore the Lord Himself will give you a sign: Behold, a virgin will be with child and bear a son, and she will call His name Immanuel."

Matthew 1:23: "Behold, the virgin shall be with child and shall bear a Son, and they shall call His name Immanuel," which translated means, "God with us."

Isaiah 44:24: "Thus says the Lord, your Redeemer, and the one who formed you from the womb, "I, the Lord, am the maker of all things, stretching out the heavens by Myself and spreading out the earth all alone..."

John 1:3: "All things came into being through Him, and apart from Him nothing came into being that has come into being."

Colossians 1:16: "For by Him all things were created, *both* in the heavens and on earth, visible and invisible, whether thrones or dominions or rulers or authorities— all things have been created through Him and for Him."

Isaiah 9:6: "For a child will be born to us, a son will be given to us; And the government will rest on His shoulders; And His name will be called Wonderful

Counselor, Mighty God, Eternal Father, Prince of Peace."

John 1:1: "In the beginning was the Word, and the Word was with God, and the Word was God."

John 1:14: "And the Word became flesh, and dwelt among us, and we beheld His glory, glory as of the only begotten from the Father, full of grace and truth."

John 5:18: "For this cause therefore the Jews were seeking all the more to kill Him, because He not only was breaking the Sabbath, but also was calling God His own Father, making Himself equal with God."

John 8:24: "I said therefore to you, that you shall die in your sins; for unless you believe that I am He, you shall die in your sins."
Note: In the Greek, "He" is not there. Thus, Jesus is saying that he is "I AM" (see John 8:58 and Exodus 3:14 passages below).

John 8:58: "Jesus said to them, 'Truly, truly, I say to you, before Abraham was born, I am.'"

Exodus 3:14: "And God said to Moses, 'I AM WHO I AM'; and He said, Thus you shall say to the sons of Israel, 'I AM has sent me to you.'"

John 10:30-33: "I and the Father are one." The Jews took up stones again to stone Him. Jesus answered them, "I showed you many good works from the Father; for which of them are you stoning Me?" The Jews answered Him, "For a good work we do not stone You, but for blasphemy; and because You, being a man, make Yourself out to be God."

John 14:6-7: "Jesus said to him, 'I AM the way, and the truth, and the life; no one comes to the Father but through Me.'"

John 14:9: "Jesus said to him, 'Have I been so long with you, and *yet* you have not come to know Me, Philip? He who has seen Me has seen the Father; how *can* you say, 'Show us the Father?''"

John 20:28: "Thomas answered and said to Him, 'My Lord and my God!'"

Acts 4:12: "And there is salvation in no one else; for there is no other name under heaven that has been given among men by which we must be saved."

Philippians 2:5-11: "Have this attitude in yourselves which was also in Christ Jesus, who, although He existed in the form of God, did not regard equality with God a thing to be grasped, but emptied Himself, taking the form of a bond-servant, and being made in the likeness of men. And being found in appearance as a man, He humbled Himself by becoming obedient to the point of death, even death on a cross. Therefore, also God highly exalted Him, and bestowed on Him the name which is above every name, that at the name of Jesus every knee should bow, of those who are in heaven, and on earth, and under the earth, and that every tongue should confess that Jesus Christ is Lord, to the glory of God the Father."

Colossians 2:9: "For in Him all the fullness of Deity dwells in bodily form."

Titus 2:13: "...looking for the blessed hope and the appearing of the glory of our great God and Savior, Christ Jesus..."

Hebrews 1:8: "But of the Son He says, 'Thy throne, O God, is forever and ever, and the righteous scepter is the scepter of His kingdom.'"
Quoted from Psalm 45:6: "Thy throne, O God, is forever and ever; a scepter of uprightness is the scepter of Thy kingdom."

2 Peter 1:1: "Simon Peter, a bond-servant and apostle of Jesus Christ, to those who have received a faith of the same kind as ours, by the righteousness of our God and Savior, Jesus Christ..."

Jesus Accepts Worship Reserved Only For God

Matthew 4:10: "Then Jesus said to him, 'Be gone, Satan! For it is written, 'You shall worship the Lord your God, and serve Him only.'''"

Matthew 2:2: "Where is He who has been born King of the Jews? For we saw His star in the east, and have come to worship Him."

Matthew 2:11:" And they came into the house and saw the Child with Mary His mother; and they fell down and worshiped Him; and opening their treasures they presented to Him gifts of gold and frankincense and myrrh."

Matthew 14:33: "And those who were in the boat worshiped Him, saying, 'You are certainly God's Son!'"

Matthew 28:9: "And behold, Jesus met them and greeted them. And they came up and took hold of His feet and worshiped Him."

John 9:35-38: "Jesus heard that they had put him out; and finding him, He said, 'Do you believe in the Son of Man?' He answered and said, 'And who is He, Lord, that I may believe in Him?' Jesus said to him, 'You have both seen Him, and He is the one who is talking with you.' And he said, 'Lord, I believe.' And he worshiped Him."

Hebrews 1:6: "And when He again brings the first-born into the world, He says, 'And let all the angels of God worship Him.'"

People Pray To Jesus

Acts 7:55-60: "But being full of the Holy Spirit, he gazed intently into heaven and saw the glory of God, and Jesus standing at the right hand of God; and he said, 'Behold, I see the heavens opened up and the Son of Man standing at the right hand of God.' But they cried out with a loud voice, and covered their ears, and they rushed upon him with one impulse. And when they had driven him out of the city, they began stoning him, and the witnesses laid aside their robes at the feet of a young man named Saul. And they went on stoning Stephen as he called upon the Lord and said, "Lord Jesus, receive my spirit!" And falling on his knees, he cried out with a loud voice, "Lord, do not hold this sin against them!" And having said this, he fell asleep."

Romans 10:9: "...that if you confess with your mouth Jesus as Lord, and believe in your heart that God raised Him from the dead, you will be saved..."

Romans 10:13-14: "...for 'whoever will call upon the name of the Lord' will be saved. How then shall they call upon Him in whom they have not believed? And how shall they believe in Him whom they have not heard?" (Paul is speaking of calling upon Jesus.)
(The phrase, "Call upon the name of the Lord," is a quote from Joel 2:32.)

1 Corinthians 1:1-2: "Paul, called as an apostle of Jesus Christ by the will of God, and Sosthenes our brother, to the church of God which is at Corinth, to those who have been sanctified in Christ Jesus, saints by calling, with all who in every place call upon the name of our Lord Jesus Christ, their Lord and ours."

(The phrase, "to call upon the name of the Lord," is a phrase used to designate prayer.)

Joel 2:32: "And it shall come to pass, that whosoever shall call on the name of the LORD shall be delivered: for in Mount Zion and in Jerusalem shall be deliverance, as the LORD hath said, and in the remnant whom the LORD shall call."

(LORD here is YHWH, the name of God as revealed in Exodus 3:14. Therefore, this quote, dealing with God Himself is attributed to Jesus).

Jesus Is The First And The Last

Isaiah 44:6: "Thus says the Lord, the King of Israel and his Redeemer, the Lord of hosts: 'I am the first and I am the last, and there is no God besides Me.'"

Revelation 1:8: "I am the Alpha and the Omega," says the Lord God, "who is and who was and who is to come, the Almighty."

Revelation 1:17-18: "Do not be afraid; I am the first and the last, and the living One; and I was dead, and behold, I am alive forevermore, and I have the keys of death and of Hades." *

* "Bible Verses That Show Jesus Is Divine." *CARM.* N.p., n.d. Web. 20 Feb. 2016. "Bible Verses That Say "Jesus Is God". N.p., n.d. Web. 20 Feb. 2016.

Appendix C: Salvation Message

Sample Prayer of Salvation

"Father, I know that I have broken your laws and my sins have separated me from you. I am truly sorry, and now I want to turn away from my past sinful life toward you. Please forgive me, and help me avoid sinning again. I believe that your son Jesus Christ died for my sins, was resurrected from the dead, is alive, and hears my prayer. I invite Jesus to become the Lord of my life, to rule and reign in my heart from this day forward. Please send the Holy Spirit to help me obey You, and to do Your will for the rest of my life. In Jesus' name I pray, Amen." *

http://www.allaboutgod.com/prayer-of-salvation.htm

If you have prayed this prayer (or one like it) earnestly and have made this proclamation of faith, then let me be the first to say, "Welcome to the family of God!" Your salvation is secure. Truly, this is the greatest decision you have ever or will ever make.

Let me also be the first to tell you that this step of faith is not just a decision of your mind, but an investment of your life. This new found belief in the existence and authority of God will radically change your life.

* *"Prayer of Salvation." AllAboutGOD.com. N.p., n.d. Web. 19 Feb. 2016.*

And since biblical faith isn't natural for us and is countercultural, you may meet opposition. I would encourage you to seek out a Bible teaching church, begin reading the Bible each day and praying, seek out a community of other Christians to whom you can be accountable, and rely on God's promises and grace to transform you into the man or woman of God that you were destined to be.

May God bless you and keep you!

References

1. VanDenBerghe, Betsy. "Spiritual IQ in a Secular Age | RealClearReligion." *Spiritual IQ in a Secular Age / RealClearReligion.* N.p., n.d. Web. 03 May 2016.

2. Augustine, Marcus Dods, John Gibb, and J. Innes, *Lectures or Tractates on the Gospel According to St. John.* Edinburgh: T & T. Clark, 1873. Print.

3. McDowell, Josh. *More Than a Carpenter.* S.1.: Living, 1973. Print.

4. Ross, Hugh, "*Reasons To Believe: Fulfilled Prophecy: Evidence for the Reliability of the Bible.*" Reasons to Believe: Fulfilled Prophecy: Evidence for the Reliability of the Bible. N.p., 22 Aug. 2003. Web. 29 Jan. 2016.

5. Ibid.

6. Strobel, Lee. *The Case for Christ: A Journalist's Personal Investigation of the Evidence for Jesus.* Grand Rapids, MI: Zondervan, 1998. Print.

7. Ross, Hugh, "*Reasons To Believe: Fulfilled Prophecy: Evidence for the Reliability of the Bible.*" Reasons to Believe: Fulfilled Prophecy: Evidence for the Reliability of the Bible. N.p., 22 Aug. 2003. Web. 29 Jan. 2016.

8. Ibid.

9. Ross, Hugh. "Reasons To Believe: Biblical Forecasts of Scientific Discoveries." *Reasons To Believe: Biblical Forecasts of Scientific Discoveries*. N.p., 1 Jan. 1976. Web. 29 Jan. 2016.

10. Ibid.

11. Missler, Chuck. *What is Truth? The Case for Biblical Integrity. Kindle Edition.*

12. Ibid.

13. Ibid.

14. McDowell, Josh. *More than a Carpenter*. S.l.: Living, 1973. Print.

15. "Jim Jones." *Wikipedia*. Wikimedia Foundation, n.d. Web. 27 Mar. 2016.

16. Rochford, James. "Evidence Unseen." *Evidence Unseen*. N.p., n.d. Web. 19 Feb. 2016.

17. Groothuis, Douglas R. *Christian Apologetics: A Comprehensive Case for Biblical Faith*. Downers Grove, IL: IVP Academic, 2011. 512.

18. Rochford, James. "Evidence Unseen." *Evidence Unseen*. N.p., n.d. Web. 18 Feb. 2016.

19. Archer, Gleason L. Encyclopedia of Bible Difficulties. Grand Rapids, MI: Zondervan Pub. House, 1982.

20. Josephus Antiquities of the Jews 20:197-203.

21. Josephus Antiquities of the Jews 4.8.15.

22. Foxe, John, and G. A. Williamson. *Foxe's Book of Martyrs*. Boston: Little, Brown, 1966. Print.

23. "World Population Since Creation." *World Population Since Creation*. N.p., n.d. Web. 28 Mar. 2016.

24. Missler, Chuck. "BEYOND COINCIDENCE - The Boundaries of Our Reality (Dr. Chuck Missler)." *YouTube*. YouTube, n.d. Web. 6 Aug. 2012.

25. Anderson, Robert. *The Coming Prince*. Grand Rapids, MI: Kregel, 1984. Print.

26. Missler, Chuck. "BEYOND COINCIDENCE - The Boundaries of Our Reality (Dr. Chuck Missler)." *YouTube*. YouTube, n.d. Web. 6 Aug. 2012.

27. Missler, Chuck. *What is Truth? The Case for Biblical Integrity. Kindle Edition.*

28. Tiegreen, Chris. One Year God with Us Devotional: 365 Daily Bible Readings to Empower Your Faith. Place of Publication Not Identified: Tyndale House, 2014. Print.

29. Ibid.

30. "Revelation - BSF International." *Revelation - BSF International*. N.p., n.d. Web. 22 Jan. 2016.

Acknowledgements

I offer my sincere thanks and gratitude to:

My daughter Olivia, for designing the front and back covers and creating the illustrations, and friend Josh Cianca for assistance in designing the back cover.

Dr. Greg Viehman, for the inspiration to write this book.

My family, friends and colleagues, Brian Gilman, Reverend Dr. Todd vonHelms, Dr. Philip Saba, Dr. M. Alan Dickens, Amy Beckman, and George Beckman for their constructive criticism and encouragement.

My wife Jill, for her wise counsel, continual support and prayers, and gentle honesty.

"...that if you confess with your mouth

Jesus as Lord, and

believe in your heart

that God raised Him from the dead,

you will be saved;"

-Romans 10:9

About The Author

Dr. William J. Vanarthos is a radiologist living in North Carolina with his wife, Jill, of 24 years. He has two daughters who are currently in college. Dr. Vanarthos enjoys reading, Bible study, working outdoors, travel, and New York Giants football.

CPSIA information can be obtained
at www.ICGtesting.com
Printed in the USA
BVHW041533210620
581975BV00006B/545